THE SCENERY OF SCOTLAND
THE STRUCTURE BENEATH

WJ...

With aerial photographs b... ...ia Macdonald

Front cover and inside front cover: **Ben Nevis**
Scotland's highest mountain is a complex layer-cake of lavas on a base of metamorphic rocks, all built up in Old Red Sandstone times. Later this whole mass foundered during cauldron subsidence. This allowed further molten rock to squeeze up through the great ring fracture created by this collapse. The hard, resistant volcanic rocks form the cliffs and buttresses of the north face.

These pictures show the mountain from the south looking over the Aonach Eagach ridge of Glencoe *(cover)* and *(opposite)* the challenging ridges of the north face leading to the summit.

Back cover: **Fields of Fife**
In this apparent agricultural Eden, southeast of St Andrews, we see the ultimate aspect of a made landscape. There is little here that has not been changed by human activity: the fields are tilled and fenced, streams straightened and trees planted.

NATIONAL MUSEUMS OF SCOTLAND

Dedicated to the late Wally Mykura, Geologist.

First published 1988 by the National Museums of Scotland, Chambers Street, Edinburgh EH1 1JF.

Third edition, 1991, reprinted 1993, 1995, 1996, 1998.

Produced by the Publications Office of the National Museums of Scotland:
    Editing: Jenni Calder
    Design: Patricia Macdonald assisted by
              Alison Cromarty.
Typeset in Ehrhardt by Artwork Associates, Edinburgh.
Printed by Clifford Press Ltd, Coventry

© Copyright Trustees of the National Museums of Scotland 1988, 1990, 1991.

ISBN 0 948636 24 6

Foreword

'That man is little to be envied, whose patriotism would not gain force upon the plain of Marathon, or whose piety would not grow warmer among the ruins of Iona.' Few would deny the enrichment which a sense of history and association brings to the traveller among scenes such as Iona, Glencoe, Glenfinnan or Strathnaver.

In the same way a new dimension is added to our enjoyment of scenery if we can see beyond its face value towards its significance as a 'still' in the motion picture of geological events. That beauty which stirs us, be it mountain, sea-cliff, strath or carse, is the result of events which have involved seas, sediments, rocks, heat, pressure, rain, wind and ice. These have wrought for us the scenery of Scotland which today exists for our wonder and tomorrow will be changed as the inexorable geological agents continue their work of destruction and creation.

This little book is an eye-opener to let us see in a new way what, for some, will be familiar and well-loved landscapes. If you have never thought before of scenery as an expression of the earth's four dimensions of structure and history, the text will set you on a road which will lead to discovery and delight. If you have already experienced the joys of that road the unique collection of aerial photographs will allow you to look at Scotland from a new and revealing angle.

Read the book in the comfort of your home and then set out to see the marvellously varied landscape of Scotland for yourself. Enjoy it with the wind in your face as the book's author and photographer have so clearly done before you.

Charles D Waterston

1 Schiehallion *(opposite)*
This sharp peak, south of Loch Rannoch, owes its form to steeply dipping Dalradian quartzites. In the foreground a delta fan is forming on the shores of the loch. Scientific experiments were conducted on Schiehallion in the 18th century in an attempt to determine the weight of the earth.

Scotland's landscape

The scenery of Scotland, although varied and beautiful, is nowhere continental in scale. There are no huge mountains or great areas of monotonous plain. Yet it is widely accepted that Scotland's scenery can be grand and magnificent in the true sense of the words, as in the Torridonian hills of the far north west. In other areas, such as the hills and sheltered fields of inland Fife, it seems a landscape in miniature.

Landscape is sculptured by the action of the weather, and in the recent geological past this also meant the grinding of debris-laden glaciers. These same glaciers left behind in many areas a mantle of debris known as *till* (see fig. 4) which can sometimes mask, but not entirely hide, the underlying rocks. The rocks beneath are like good bone structure in an old lady's face. Even in old age they allow the landscape to maintain both beauty and character.

4 Quarry, south of Callander *(right)*
This quarry has been dug to allow the extraction of sand and gravel from glacial deposits. These mounds of debris on a layer of the underlying hard bouldery clay called *till* are the deposits of a glacier at the end of the last ice age. This type of scenery, known as 'basket-of-eggs', is characteristic of many glacial deposits.

2 Torridonian landscape *(opposite)*
The sandstones of the Torridonian were laid down on an already ancient Lewisian landscape in Precambrian times. Borne along by the mighty rivers of a continent lying to the northwest of Scotland a mass of debris was deposited to a depth of 7 km in some places. Then the inevitable processes of erosion began to wear it away. All that now remain are the relict mountains of the Torridonian sandstone, like Suilven in Assynt, rising above the barren moorland in sweeping buttresses and serrated knife edges.

3 Fife *(right)*
The small hills, little valleys and delicate landscape of Fife reflect the geology. The large number of small volcanic features such as vents, plugs and sills set in a ground mass of Carboniferous and Old Red Sandstone sedimentary rocks has been picked out by ice. The differences in rock type are emphasized by present-day farming practices and land use.

Rocks are the skeleton of the landscape and they lay down the limitations and boundaries for its final aspect. The Lewisian rocks (named after the island of Lewis) of the ancient basement have familiar characters which would allow those who have studied them to be blindfolded, flown three times round the world, and if landed on a landscape of Lewisian rocks to say 'I may not know exactly where I am but these are rocks of the Lewisian'. Each rock type has its own indicator or group of indicators which show through in the landscape. This may be the greenness of the vegetation over limestone, or the predominance of heather over granite, or a physical indication such as the table-like feature created by Tertiary basalts (see figs. 5, 6 & 7).

5 Lewisian Gneiss
Almost 3,000 million years old, the Lewisian Gneiss is a complex of the most ancient rock types in Britain and is one of the oldest in the world. Its tough resistant rocks form a distinctive scenery in the northwest Highlands and the Outer Hebrides. The gneiss forms a low undulating plateau showing an endless succession of small hills and ridges of bare rock, among which lie innumerable lochans.

Scotland's landscape

Scotland's landscape has been built up from a great variety of rock types, each of which has had its own effect on the surface that we see. Many eminent geologists have taken hundreds of pages to describe the structure and foundation of the landscape, and the work of pioneering geologists in Scotland is reflected in the use of many Scottish placenames for important rock types and features. As this is only a short introduction to Scotland's geology and scenery, it can only be a very simplified outline. For those readers who wish to find out more, there is a book list at the back.

6 Lismore *(above)*
This island in the Firth of Lorne is formed completely of Dalradian limestone. Once known as the garden of Argyll because of its thin but fertile soil many of its fields are now rushy and neglected. However, in spring and summer its bright green grass shows up like an emerald against the darker highlands.

7 Healabhal Mhor (Macleod's Table, North), Skye *(right)*
This little plateau, one of a pair southwest of Dunvegan, Skye, has been weathered from the almost horizontal layers of basalt lavas. It is said to be on one of these that a chief of the Macleods held a great feast for the King. With the stars for a ceiling, his clansmen as torch bearers and the green grass as a table cloth he showed that his claim at a royal banquet that he had a larger and finer banqueting hall was not just an idle boast.

The geological foundations

The geology of Scotland is very complex, and although it lacks certain elements common on the Continent, such as sediments of Tertiary age, active volcanoes and glaciers, it is on first acquaintance a formidable mixture of rock types and chronological periods. Fortunately, with some exceptions, it can be broken down into four geographical areas. These are:
1. The Northwest Highlands and Islands: the basement area of the Outer Hebrides and the Highlands west of the Moine Thrust.
2. The Northern and Central Highlands, between the Moine Thrust and the Highland Boundary Fault.
3. The Central Lowlands south of the Highland Boundary Fault.
4. The Southern Uplands Fault to the English border.

8 Satellite photograph of Scotland *(ERSAC)*
Landsat mosaic in simulated natural colour. A simple interpretation of the colours is given below:

Agricultural crops	Strong green
Less developed crops & grass	Lighter green
Woodland	Very dark green
Heather moorland	Dark brown
Peat moorland	Greyish brown
Grassland	Light brown
Bare soil	Brownish white
Cities and towns	Grey
Coastal sand/muds	Grey/white
Water	Dark blue
Snow	White

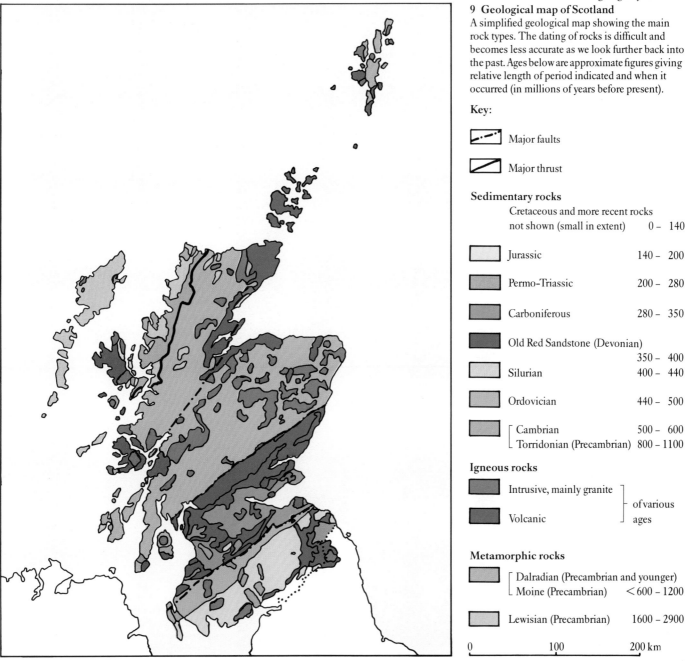

The geological foundations

9 Geological map of Scotland
A simplified geological map showing the main rock types. The dating of rocks is difficult and becomes less accurate as we look further back into the past. Ages below are approximate figures giving relative length of period indicated and when it occurred (in millions of years before present).

Key:

⟋·⟍ Major faults

⟋   Major thrust

Sedimentary rocks

Cretaceous and more recent rocks not shown (small in extent)   0 – 140

Jurassic   140 – 200

Permo-Triassic   200 – 280

Carboniferous   280 – 350

Old Red Sandstone (Devonian)   350 – 400

Silurian   400 – 440

Ordovician   440 – 500

Cambrian   500 – 600
Torridonian (Precambrian)   800 – 1100

Igneous rocks

Intrusive, mainly granite ⎱ of various
Volcanic ⎰ ages

Metamorphic rocks

Dalradian (Precambrian and younger)
Moine (Precambrian)   <600 – 1200

Lewisian (Precambrian)   1600 – 2900

0   100   200 km

The geological foundations

1 The Northwest Highlands and Islands

This area for convenience should also include the Tertiary volcanics of the Inner Hebrides. Four major rock types each give rise to a significant land form. Lewisian rocks, the oldest in Britain, are a re-exposure of an ancient landscape. Rarely forming hills of more than a few hundred feet high, they produce a knobbly terrain of bare rock knolls between pools of peaty water. The surfaces of the knolls are often rounded by ice and the whole effect is one of a chaotic, and yet repetitive, pattern of rock and water.

The Torridonian lies directly upon the Lewisian but is now only a remnant of a once great expanse. The almost horizontally bedded warm red, pebbly sandstones form relict mountains which rise upon their Lewisian foundations like the ruined castles of Elfland. No one who has seen the Torridonian mountains will need to be convinced of their special attraction and grandeur. Anyone who has not seen them should do so.

10 Manish, Harris
Major striations and fracture lines have been emphasized by weathering, which exposes the 'grain' of the rock.

The igneous rocks of this area are of two types. (Look at fig. 55, page 35 for an explanation of 'igneous' and other terms.) First are extrusive lavas such as those which create the basalt plateaux of Northern Skye (fig. 12). The second are intrusive complexes which result in the spectacular scenery of such places as St Kilda and Rhum, and in Skye are known to mountaineer and photographer in the jagged gabbro peaks of the Black Cuillins (see figs. 13 & 14 overleaf) and the rounded granite mountains of the neighbouring Redhills.

11 Meall Mheadhonach, east summit of Suilven
The castellated summit of Suilven shows the almost horizontal bedding of the Torridonian sandstone. The landscape seems eternal, but in the few thousand years since the melting of the last ice sheet a new phase of weathering has begun.

12 Basalt cliffs at Geary, Skye
The Tertiary basaltic lavas of northern Skye erupted in a very fluid state. They spread quickly before hardening to form a series of shallow and almost flat layers. Breaking down to form rich soils they have always been important for crofting. At Geary ruined crofts and walls show that farming has been going on here for a long time.

The geological foundations

13 & 14 Cuillins

In the Tertiary igneous centre of the Black Cuillins on Skye we see what for many is the most exciting mountain scenery in Britain. The gabbro which forms their peaks has been eroded by the elements to form narrow ridges and pinnacles whose summits are only trodden by skilled climbers. Even to gaze into the glaciated hollow of Loch Coruisk lying in its ice-carved niche requires either a boat journey or a long walk by the Bad Step.

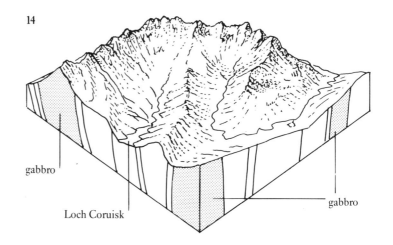

14

15, 16 & 17 Quiraing and Storr, Skye

The spectacular basalt lava flows of northern Skye have formed a scenery of plateau features which make up the inland hills and great coastal cliffs. At the Quiraing *(lower right)* and the Storr *(below)* these cliffs have been destabilized, probably by the retreat of the ice from the Sound of Raasay at the end of the last ice age. This has resulted in the break-up of the weaker Jurassic sedimentary rocks underlying the lavas, causing massive landslides which have torn and twisted the rocks into weird shapes and distorted the very surface of the earth. The diagram *(upper right)* shows a simplified section of the Quiraing, illustrating this process.

The Northwest Highlands and Islands

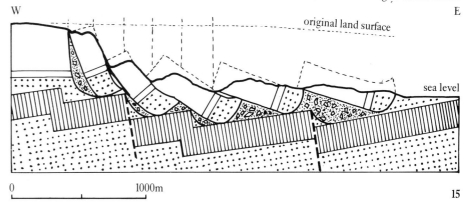

15

The geological foundations
2 The Northern and Central Highlands

Again we can talk of four main rock types, although none will give rise to such unique features as in the previous area. Predominant are ancient metamorphic rocks of the Moine and Dalradian overlain in the north and east by younger sedimentary rocks. Also included are the granite batholiths which colour the geological map red and provide the clear burns for the whisky distilleries.

The metamorphic rocks were originally deposited as sediments but rendered crystalline by heat and pressure, folded during the Caledonian mountain-building era and finally carved and sculpted by glacial ice.

Even a landscape enthusiast may find the flaggy quartzites of the Moine in the Northern Highlands occasionally boring whereas the more varied geology of the Dalradian in the Central Highlands provides a rare seasoning of limestone schists which can make an alpine garden of a Perthshire mountain.

18 Glencoe
The complex rocks of the Glencoe mountains are the eroded remnants of volcanic activity in Old Red Sandstone times. During a period of eruptions the magma chamber underlying the Glencoe area was almost emptied and the reduced pressure caused the rocks above to founder, creating a giant ring fracture. Molten rock welled up from this great crack to form a ring dyke. This feature can be picked out by studying the various rock types exposed in the great cliffs and looming buttresses of this impressive glen.

The summit of Ben Nevis may be seen in the distance.

The Northern and Central Highlands

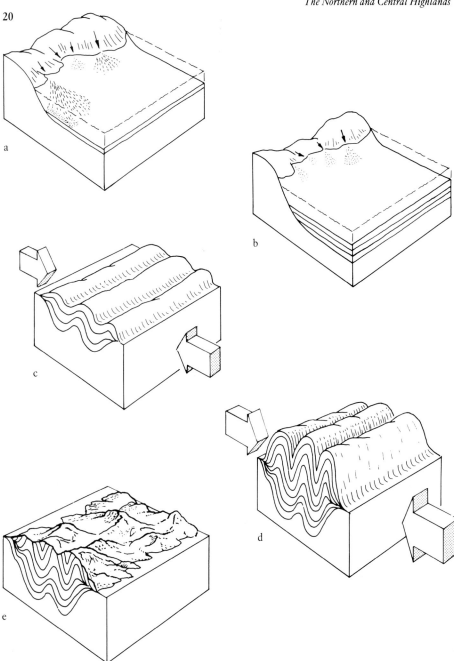

19 & 20 Folding

Sediments are laid down in layers often thousands of feet thick. As the bottom of the basin, where sedimentation is taking place, continues to sink more layers are deposited on top, and the deposits get buried deeper and deeper. The increasing weight causes extreme pressure on the bottom layers so that water is squeezed out and the soft material left is cemented into solid rock.

During a period of mountain building the forces acting near the earth's surface are extremely powerful. The rocks are pushed and squashed from different directions and begin to rise and fold. They become altered in the process – how much depends on the pressure and temperature. These altered sediments, called metamorphic rocks, are usually harder than the surrounding sediments. Folded rocks when exposed at the surface are quickly eroded by glaciation and weathering. This causes changes in the surface shape and wears through to expose underlying layers and features.

Ben Lui *(above)* is a mountain created by these processes.

The geological foundations

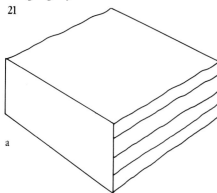

21 How slate is formed *(left)*
Layers of clay are buried deep within the earth, where they are compressed and hardened. Pressure from the sides causes the layers to buckle and fold. This begins to recrystallize the clay minerals. When the layers have folded the recrystallization is complete. The crystals, mainly mica, which is a shiny, flaky mineral, lie parallel to each other and at right angles to the direction of pressure. It is along these planes of weakness (cleavage planes) formed by the mica crystals that slate can be split.

22 Flooded slate quarry, Belnahua, Firth of Lorne *(right)*
Roofing slates were worked on this tiny island from at least as early as the 18th century. This quarry had reached a depth of over 90 feet, with only a thin outer shell of black slate between it and the sea, before commonsense prevailed and working ceased in the early part of this century. Notice how the outlying islets are parallel to the cleavage plane of the slate. The now flooded quarries indicate the original workings along that cleavage.

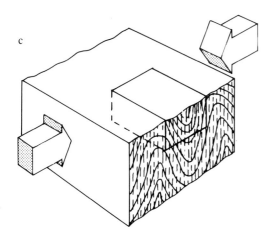

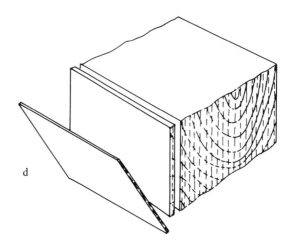

23 & 24 Ardnamurchan

The Tertiary volcanic area of Ardnamurchan is very complex. It consists of a series of cone sheets, which were intruded when pressure in the magma chamber was high, and a series of concentric dykes intruded into the surrounding rocks when the roof of the magma chamber collapsed. Cone sheets are narrow, cutting the country rocks at an angle which goes inwards and downwards towards the volcanic centre, while ring dykes are thicker and angle outwards as they go down. Deep erosion since Tertiary times has now exposed the ring dykes as curved ridges in the landscape.

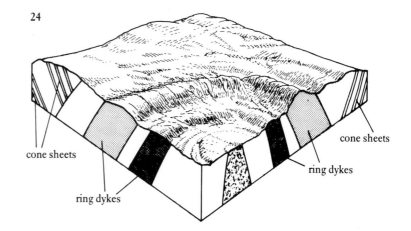

The geological foundations

Overlying the eastern and northern edge of the eroded Caledonian fold belt are the sedimentary rocks of the Old Red Sandstone and the Jurassic. The flaggy sandstones of Caithness and the Orkney islands are the uplifted freshwater sediments from the bed of 'Lake Orcadie'. Inland they form low rolling hills but on the sea coast spectacular cliffs and other extraordinary erosional phenomena such as geos, gloups and sea stacks.

The Jurassic sediments are coastal marine in origin and of very limited exposure on the east coast. Best seen between Golspie and Helmsdale, it is possible to find here the remains of ancient coral reefs and to stand facing the sea as it was in Jurassic times, and in one's mind's eye to see ichthyosaurs leap from the warm ocean and imagine the roar of dinosaurs in the dense jungle behind.

The Northern and Central Highlands

25 Flotta *(opposite left)*
This is a typical example of the low-lying Orkney islands whose underlying rock is the flaggy Middle Old Red Sandstone. The dead flooded quarry in the foreground is a reminder of the economic geology of the past. The distant oil terminal illustrates the development of newly discovered resources.

26, 27 & 28 Stacks and other sea-cut features, Hoy, Orkney *(this page and opposite right)*
The formation of stacks is an awesome reminder of the power of the sea. Taking advantage of those joints in the rock which lie at right angles to the shore the sea first cuts a cave which may have a gloup or blow hole at its landward end. Eventually the roof of the cave collapses leaving a narrow channel or geo. Then currents swirl round to pick another joint at right angles to the first. The process is repeated until eventually a great stack like the Old Man of Hoy stands for a few thousand years, until the sea eats away the foundations and brings it crashing down.

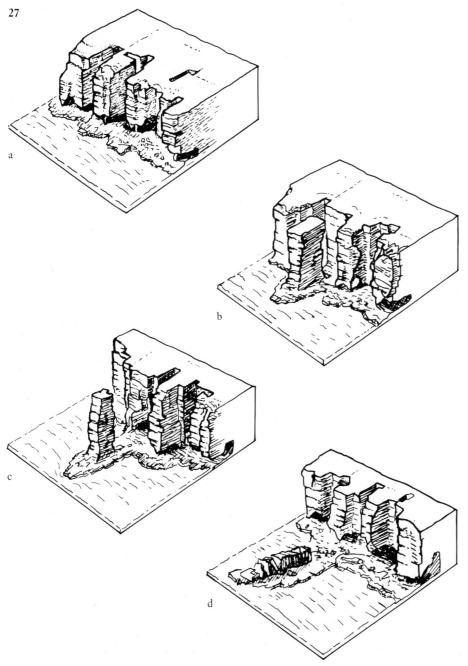

The geological foundations

From scenes such as these we change in an instant to the Arctic plateau of the Cairngorms. This granite batholith was intruded at depth into the surrounding rocks during Devonian times and, as the overlying rocks became eroded, it gradually became the granite massif we see today. The rock, although a source of treasured crystal, is poor in nutrients and its high average height produces winter snowfields and weather as though it were 1000 km further north. A combination of steep sided corries, glaciated valleys and rounded surface plateaux with occasional tors, can give the Cairngorms an interesting but innocuous look during good weather. It is as well to remember however that the scale of the mountains is greater than anywhere else in Britain and the weather changes can be sudden and dangerous.

29 Cairngorms *(above)*
The Cairngorm mountains are the remains of an immense mass of granite intruded deep within the country rocks about 400 million years ago. Since then continual movement and erosion of the rocks above have brought the granite to the surface to form Scotland's only arctic environment on the plateau surrounding the Braeriach and Ben Macdhui summits. The homogeneous nature of the granite rocks is broken in places by veins of the more massively crystalline pegmatite, which in occasional cavities contains rare and beautiful crystals such as topaz, beryl, morion and sherry-coloured cairngorm. During Victorian times miners worked the veins for these rich pockets with hammer, chisel and black powder, living in rough stone shelters beside their claims.

30 Lochan Uaine, Cairngorms *(left)*
A shallow lochan occupies the ice-eroded hollow in this high corrie on the Angel's Peak in the Cairngorms. The mountain is said to have been so named by the climber Alexander Copland to keep the nearby and lower Devil's Point in its place.

31 & 32 U-shaped valleys *(opposite)*
Passage of a glacier down an existing river valley reshapes it completely. Carrying with it a great mass of rock debris it acts as a giant file, grinding and smoothing as it moves relentlessly on. As can be seen in the photograph, of Glen Dee at the southern end of the Lairig Ghru, the action of the ice cuts off spurs from the valley sides, makes the floor flatter and the sides steeper. In cross section the valley has a typical U shape.

The Northern and Central Highlands

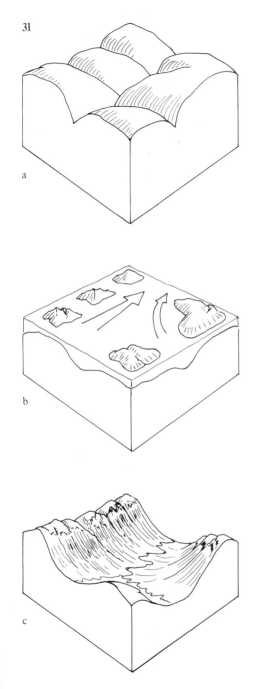

31

a

b

c

21

The geological foundations

34 Map showing the major faults

The Moine Thrust
The Moine Thrust is a special kind of fault where one layer of rock has been pushed over another.

33 Great Glen Fault *(left)*
Fractures take place along weak lines in the earth's crust, allowing blocks of rock to move against each other. Some faults allow movement in a sideways direction whilst others cause one block to push upwards and the opposing block to be pushed down. The Great Glen Fault is a sideways tear in the earth's crust that has been much deepened by ice during the last glacial period.

The Northern and Central Highlands

35 & 36 Highland Boundary Fault *(this page)*
Stretching in a diagonal line across Scotland from the Firth of Clyde in the southwest to Stonehaven in the northeast this fault separates the Highlands to the north from the Lowlands to the south. It is one of the most important and obvious geological features in Scotland. It is normally most evident where the rocks of the Midland Valley to the south have moved downwards against the Highland rocks, leaving a step-like feature in the landscape. Only occasionally is the fault itself represented, where the intruded and upturned rocks have resisted weathering and, as here at Loch Lomond, show as a ridge in the landscape.

The Southern Uplands Fault complements the Highland Boundary Fault and forms the southern edge of the Midland Valley.

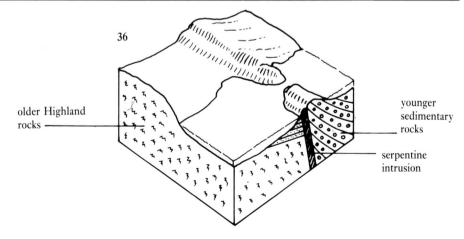

3 The Central Lowlands

In this area lies the biggest concentration of the economically valuable rocks. Carboniferous in age these rocks have been mined and quarried for centuries. It is really only in the central lowlands that the hand of modern man can be said to play a major part in the physical shaping of the landscape. If we visualize the central lowlands as a trough which was gradually infilled by debris derived from higher areas to north and south, then intruded and overlain by volcanic episodes, a simplified picture of its origin begins to emerge. Mainly appearing at the surface on the north of the trough are the conglomerates and sandstones of the Old Red Sandstone. Rich in iron, they colour the fertile fields of Tayside, Angus and parts of East Lothian a dark red.

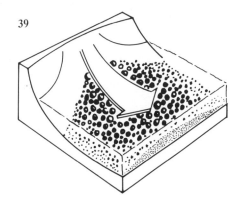

38 Meanders of the River Forth, near Gargunnock *(above)*
The power of rivers on rock is shown by the alluvial plains they lay down in their lower reaches. In the valley of the Forth the river twists and turns as it flows slowly seaward down the gentle gradient.

39 Conglomerate or puddingstone *(left)*
This rock was formed from sand, pebbles and boulders washed down hills in an ancient landscape. As the speed decreased at the bottom of the slope so its carrying capacity was reduced. This resulted in the larger boulders being dropped, while smaller pebbles were carried on until they also were too heavy for the water to carry. Larger pebbles and boulders when cemented together form conglomerate.

37 Red soil near Garvald *(opposite)*
The newly ploughed field at the bottom of this picture is the beautiful red colour of soils derived from the Old Red Sandstone. Stained by iron minerals these soils are not only colourful but fertile. They can be seen here in East Lothian, and also in Perthshire and Angus.

The geological foundations

In Carboniferous times, shallow seas and great deltas laid down a cornucopia of rock types which gave rise to the wealth of the region and to the concentrated effect of man upon its scenery. Limestone and coal, oil shale and ironstone, sandstone and clay: they have been mined, quarried, processed and dumped until the landscape has been altered so much that we are no longer aware of its original face. In some cases, such as the burnt oil shale heaps, the residues themselves have become part of the new landscape.

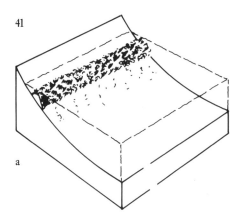
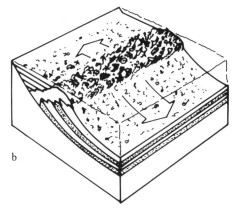

41

a
b

40 Stone quarry near Stirling *(below)*
A quartz dolerite sill being worked for roadstone.

41 & 42 Limestone
Some limestones are formed by the shells of small sea creatures which lived in reef communities. They took calcium and carbon dioxide dissolved in sea water to provide the calcium carbonate for their shells. After death the hard material dropped to the sea bed and was washed to either the inner or outer reef, gradually collecting on the bottom. In time it formed a soft layer of calcium carbonate which later hardened into limestone *(see above)*.

Among its various economic uses limestone is the basic material for cement. The cement works illustrated below is near Dunbar.

43 Shale bings *(above)*
Great red mounds of 'cooked' shale are now all that remain of Scotland's once important oil shale industry. Lower Carboniferous shales were mined in West Lothian. The production of oil, obtained by the destructive distillation of these shales, reached a peak early this century. The industry ceased in the early 1960s, when cheap oil from the Middle East made further production uneconomic.

44 Glensanda *(right)*
The scars in the foreground are but a small indication of the giant quarry in the Strontian granite that is being carved out behind them. This quarry, an example of large-scale extraction, will provide high-quality stone loaded directly into ships anchored close inshore in the ice-deepened waters of Loch Linnhe. Started to meet demands in the south of England, Europe and the USA it shows that economic quarrying occurs where a suitable source exists, even at a distance from the end user. The quarry may eventually become the biggest in Scotland.

The geological foundations

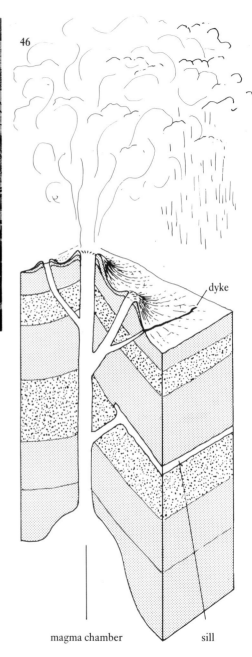

There are numerous complex and varied volcanic features in the central belt ranging in age from Old Red Sandstone to Permian and in size from the Ochil Hills to single volcanic necks such as the Bass Rock. The relationship of these rocks to their surroundings is often further complicated by faulting and subsequent glacial activity. The scenery associated with the Ochils fault, for example, is a combination of volcanic activity, sedimentary rocks, faulting, glacial action, erosion and deposition of alluvial fans, subsequently built on by man. However, with careful observation and a good guide book, the resulting landscape can be understood.

45 Arthur's Seat *(above left)*
The sleeping lion overlooking Edinburgh is a reminder of the area's ancient volcanic past. In Carboniferous times five vents spewed out dust, ashes and at least 13 lava flows. The prominent feature in the foreground of the photograph is a large sill, the outcrop of which forms Salisbury Crags. Subsequent to the volcanic activity, tilting and erosion took place. Now all is peaceful and the eroded remains provide ideal field examples for budding geologists.

46 Simplified diagram of a volcano *(above)*

The Central Lowlands

47 Bass Rock
The steep cliffs of the Bass, off the East Lothian coast, are the sides of a congealed mass of liquid rock which hardened in the throat of an ancient volcano. This is known as a volcanic plug. Now the phonolite rock is home to a multitude of gannets, which swirl around the Bass like snowflakes.

4 The Southern Uplands

The Southern Uplands fault delineates the northern boundary of the band of hills that extends to the English border. Mainly of Ordovician and Silurian sediments, this area too is variegated by patches of Carboniferous and New Red Sandstone sediments and by the major intrusive granite masses of Criffel, Cairnsmore of Fleet and Loch Doon. To the superficial observer the repetitive green hills of the Southern Uplands may look like the waves on a green ocean rolling ever southwards to crash at last on the northern bastions of England. However, to the Borderer and keen observer alike they are as varied as the geology. On the rocky coast at Ballantrae, great lava pillows are piled on the cliffs looking as fresh as they must have done millions of years ago. On the Berwickshire coast the intense folding of the rocks is shown in the magnificent cliff exposures of Fast Castle and Heathery Carr.

49 Fast Castle *(right)*
Perched on a jutting cliff above the North Sea, Fast Castle in Berwickshire has a long history. It was a point of conflict between Scots and English in the 15th and 16th centuries, and is also the 'Wolf's Crag' of Sir Walter Scott's *The Bride of Lammermoor*. Its geological history is much older, as it stands upon the steeply folded greywackes, mudstones and siltstones of the Silurian system.

50 Southern Uplands *(below)*
This entire rather uniform region consists mainly of sedimentary rocks belonging to the Ordovician and Silurian systems. There is a regularity to their geological structure which consists of a series of folds running northeast to southwest. Nobody could describe the rocks better than Sir Archibald Geikie who said, 'They have been crumpled up like a pile of carpets'.

48 Eildon Hills *(opposite)*
These are the denuded remains of a series of mushroom-shaped bodies of lava intruded into Upper Old Red Sandstone rocks during Carboniferous times. The sedimentary rocks which once covered these lavas have now been totally weathered away, leaving the softer rocks to lap round the bottom of the hills like the sea against an island shore. This was a favourite view of Sir Walter Scott, who lived nearby at Abbotsford on the River Tweed.

The Southern Uplands

In between are rounded hills and deep valleys as potent in the ability to free the spirit as any mountain top in Glencoe. Glacial action has here had a more rounding effect than anywhere else in Scotland, culminating in such classic features as the deepened valley of the Devil's Beeftub and the wide dome of the summit of Broad Law.

Although the predominant rocks are of the Lower Palaeozoic the granite masses of the southwest give a sense of being back in the northern Highlands. Small in extent, but unique in character, the New Red Sandstone rocks are petrified sand dunes now supporting green grass and dairy cows.

51 & 52 Siccar Point

The almost vertical rocks at the bottom of the cliffs of Siccar Point, Berwickshire *(opposite)* are covered by younger rocks which dip gently seaward. The unconformity which occurs at the boundary between these rocks makes this a place of geological pilgrimage. The lower rocks, called greywackes, are much older than those above and belong to the Silurian system. They were already folded and partly weathered away before the deposition of the Old Red Sandstone conglomerates which now overlie them. *(See diagrams on this page.)*

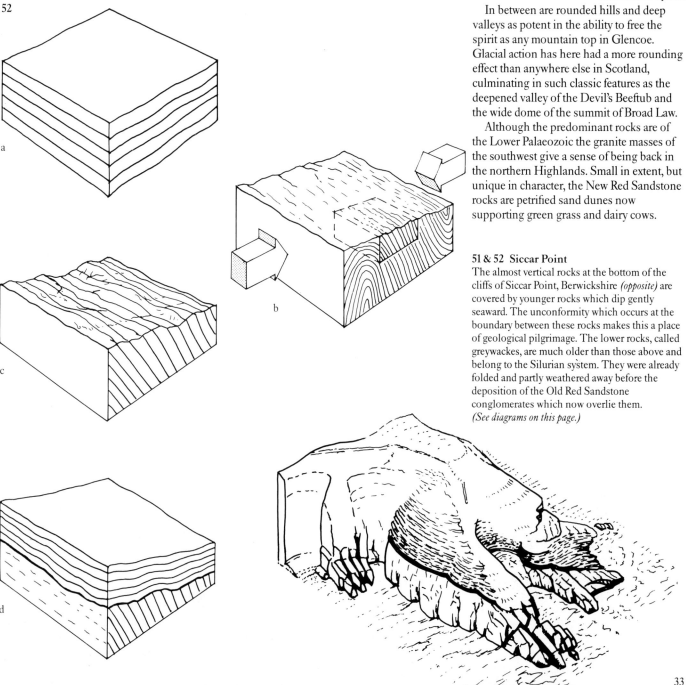

A closer look

This is a simple breakdown of the geological foundations of Scotland, but it provides the observer with a basis on which to add more detail. For instance, the many volcanic dykes (see fig. 53) which cut across the various rock types can be picked out as ridges or gullies, depending on their hardness relative to that of the rock through which they pass. Small areas of ultrabasic rocks give bright spots of green, which is not that of limestone, but the *achadh uaine*, the green field, of the crofter. Faults such as the Great Glen traverse the landscape creating boundaries and pathways, while mineral veins (fig. 54) show up as streaks and slashes depending on whether they are barren quartz or have already been mined for their ore.

Each area is unique and varied, the differences outweighing the similarities. For those who learn to read the rocks, the scenery will be a never-ending wonder.

53 Dykes *(above)*
During periods of volcanic activity molten rock can sometimes be pushed upwards through fractures in the overlying rocks. The molten rock soon solidifies and if later the surrounding rocks are worn away the intruded rock can be left standing upright on the landscape, as here at Eilean Duin, Firth of Lorne.

Change and conservation

The conservation of the landscape is a subject guaranteed to raise friction and controversy. The pock-marked outer skin is likely to be man's work. The deeper layers are all the province of the rocks beneath.

Those whose life and work involves changing the face of our landscape should have at least a basic knowledge of its anatomy and skeletal framework. This applies not only to miners and quarrymen, but to farmers and foresters, to civil engineers and hydrologists, sewage and rubbish disposal operatives, and last and perhaps most important of all, the general public. It is in their name and often with public money that projects which will alter the landscape are funded. Changes to the landscape tend to be permanent, at least on the human scale of time, so they should not be carried out thoughtlessly or carelessly. This is the only earth we will ever stand on and although it can be given a face lift, it can never be traded in for a new model.

While people now have the power to damage and even destroy a landscape they have only the gardener's ability to create one. However, Scottish landscape is of such quality and variety that there still exists a multitude of beautiful scenes from coast to coast and roadside to mountain top. Neither an open book nor an unfathomable mystery, its genesis can be understood easily by those who hold the key to its veiled dimension, the structure beneath.

54 Strontian lead mines *(below)*
Lead-bearing veins were discovered in Argyll in the 18th century. Trending east-west for over 7km they occupy a shear zone on the north side of the Strontian granite near the contact with Moine rocks. The orebody contains barite, calcite and quartz with sphalerite and silver-bearing galena. Other minerals are strontianite and the rare zeolites brewsterite and harmotome. The element strontium was first isolated from material found here and named after it.

Initially the vein was mined for lead and silver and, although closed for many years, it is now being worked again, mainly for barite.

Types of rock

Rocks can be grouped either by age or by mode of origin. Those mentioned in the text and captions are grouped below according to their mode of origin. In the booklet rocks are also referred to by names that indicate age, eg Carboniferous. (See the geological map, fig. 9.)

Igneous rocks
These can be regarded as the primary source of materials comprising the earth's surface. They mostly solidified from a liquid state and often have a crystalline appearance.

There are two types of igneous rocks: *extrusive*, which poured out on the surface, and *intrusive*, which intruded into overlying rocks.

Extrusive:
Ash
Lavas (basalt and phonolite are types of lava)

Intrusive:
Gabbro
Granite
Pegmatite

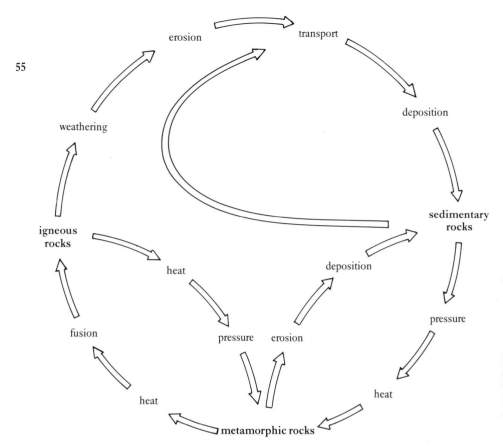

Sedimentary rocks
These are formed from material either derived from pre-existing rocks by weathering, or composed mainly of particles with an organic origin.

From weathering:	*Mainly from organic sources:*
Clay	Coal
Conglomerate or puddingstone	Limestone
Greywackes	Oil Shale
Ironstone	
Mudstones	
Sandstone	
Shale	
Till	

Metamorphic rocks
These are rocks whose nature has been changed by heat and pressure within the earth's crust.

Gneiss
Quartzite
Schist
Slate

55 The Rock cycle *(above)*
The continuous processes of the rock cycle show in a simple way how rocks can be built up, altered and destroyed. All the stages will be happening in different places, now. The stages of the cycle are not fixed and need not be taken to completion. Sands deposited one day may be washed away by a flood the next, and igneous rocks intruded at depth may never show themselves at the surface before being consumed again within the earth.

Bibliography

Please note that some titles are now out of print, but should still be available in libraries.

Alison, I et al. 1988. *An excursion guide to the Moine Geology of the Scottish Highlands.* Edinburgh: Scottish Academic Press.

Barber, A J et al. 1978. *The Lewisian and Torridonian rocks of North West Scotland.* Guide no. 21. London: Geologists Association.

Bell, B R and Harris, J W. 1986. *Excursion Guide to the Geology of Skye.* Glasgow: Geological Society of Glasgow

Black, George P. 1966. *Arthur's Seat.* Edinburgh: Oliver & Boyd.

Bluck, B J (ed.) 1973. *Excursion Guide to the Geology of the Glasgow district.* Glasgow: Geological Society of Glasgow.

British Geological Survey. Regional Geology of Scotland, covering Grampian Highlands, Northern Highlands, South of Scotland, Midland Valley of Scotland and the Tertiary Volcanic Districts. Obtainable from government bookshops or British Geological Survey, Murchison House, West Mains Road, Edinburgh.

British Geological Survey. Memoirs of The Geological Survey of Scotland. Obtainable from the same address for some, but not all, of the old Scottish counties. They cover a smaller area in greater detail.

Brown, G M. 1969. *The Tertiary Igneous Geology of the Isle of Skye.* Ed J G Capewell. London: Geologists Association.

Clarkson, E N K and McAdam, A D (eds). 1986. *Lothian Geology, an excursion guide.* Edinburgh: Scottish Academic Press.

Craig, G Y (ed.) 1990. *The Geology of Scotland.* 3rd ed. Edinburgh: Scottish Academic Press.

Emeleus, C H. 1979. *Field guide to the Tertiary igneous rocks of Rhum, Inner Hebrides.* (Peterborough): Nature Conservancy Council.

Geikie, A. 1865. *The Scenery of Scotland.* 1st ed. London: Macmillan & Co.

Gray, J M and Lowe, J J. (eds). 1977. *Studies in the Scottish late glacial environment.* Oxford, Pergamon Press.

Gribble, C D, Durrance, E M and Walsh, J N. 1976. *Ardnamurchan: a guide to geological excursions.* Edinburgh: Edinburgh Geological Society.

Johnson, M R W and Parsons, I. 1979. *Assynt District of Sutherland.* Edinburgh: Edinburgh Geological Society.

Lawson, James and Lawson, Judith. 1976. *Geology explained around Glasgow and south-west Scotland including Arran.* Newton Abbot: David and Charles.

McCallien, W J. 1938. *Geology of Glasgow and district.* London: Blackie & Sons Ltd.

MacDonald, J G and Herriot, A. 1983. *Macgregor's excursion guide to the geology of Arran.* 3rd ed. Glasgow: Geological Society of Glasgow.

MacGregor, A R. 1968. *Fife and Angus Geology: an excursion guide.* London and Edinburgh: William Blackwood & Sons Ltd for the University of St Andrews.

Poucher, W A. 1980. *Scotland.* London: Constable.

Poucher, W A, 1983. *The Highlands of Scotland.* London: Constable.

Price, Robert J. 1976. *Highland landforms.* Inverness: Highlands and Islands Development Board.

Roberts, John L. 1989. *Geological structures.* London: MacMillan.

Sissons, J B. 1967. *The evolution of Scotland's scenery.* Edinburgh: Oliver & Boyd.

Steers, J A. 1973. *The coastline of Scotland.* Cambridge: Cambridge University Press.

Trewin, N H et al. 1987. *Geology of the Aberdeen area.* Edinburgh: Scottish Academic Press.

Upton, B G J. n.d. *Carboniferous volcanic rocks of the Midland Valley of Scotland.* Edinburgh: Edinburgh Geological Society.

Walker, Frederick. 1961. *Tayside geology.* Dundee: Dundee Museum and Art Gallery.

Walker, Frederick. 1963. *The Geology and scenery of Strathearn.* Dundee: Dundee Museum and Art Gallery.

Whittow, J B. 1979. *Geology and scenery in Scotland.* With revisions. Harmondsworth: Penguin Books Ltd.

Guides published by the Edinburgh Geological Society can be obtained from the Publications Officer, Edinburgh Geological Society, c/o The Grant Institute of Geology, King's Buildings, West Mains Road, Edinburgh.

Acknowledgements

All the diagrams were produced by Michael Spring with his customary enthusiasm and patience, some based on originals by Alan Chalmers of the Forestry Commission.

Satellite imagery in fig. 8 courtesy of the Environmental Remote Sensing Applications Centre, Livingston, West Lothian (Tel: 0506 412000).

The author and photographer are also grateful to the following: staff of the British Geological Survey, Edinburgh and of the Department of Geology, National Museums of Scotland, Ian Bunyan, Kathleen Davidson, Angus Macdonald, Dr Charles Waterston and Colin Will.

Inside back cover: **Westfield opencast site.** At opencast sites like this one in Fife coal is obtained by quarrying. With the use of giant earth-moving machines the debris overlying coal seams near the surface is removed and the coal extracted. It looks at present like a great ulcer in the landscape. But when work is finished and the hole is filled in the surface can be returned to agriculture or used for building. With sensitive restoration all surface traces of the great pit will vanish.